Nature of Life

Animal Cells

Michelle Lomberg

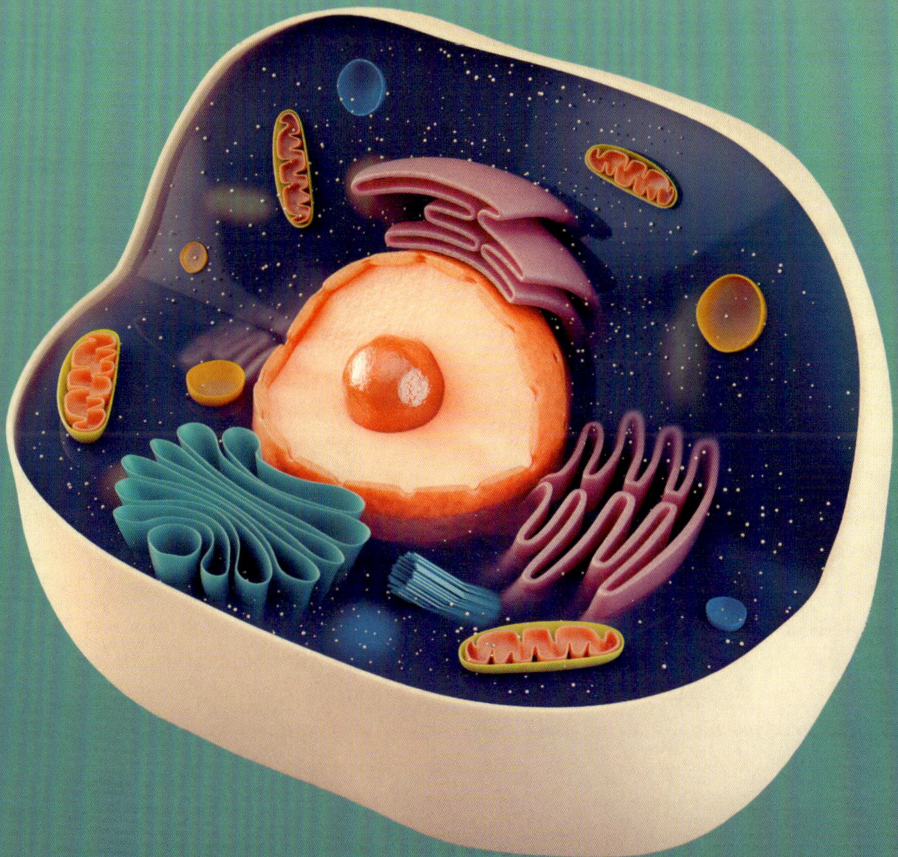

LIGHTBOX
openlightbox.com

LIGHTBOX

Go to
www.openlightbox.com
and enter this book's
unique code.

ACCESS CODE

LBM48828

Lightbox is an all-inclusive digital solution for the teaching and learning of curriculum topics in an original, groundbreaking way. Lightbox is based on National Curriculum Standards.

STANDARD FEATURES OF LIGHTBOX

AUDIO High-quality narration using text-to-speech system

ACTIVITIES Printable PDFs that can be emailed and graded

SLIDESHOWS Pictorial overviews of key concepts

VIDEOS Embedded high-definition video clips

WEBLINKS Curated links to external, child-safe resources

TRANSPARENCIES Step-by-step layering of maps, diagrams, charts, and timelines

INTERACTIVE MAPS Interactive maps and aerial satellite imagery

QUIZZES Ten multiple choice questions that are automatically graded and emailed for teacher assessment

KEY WORDS Matching key concepts to their definitions

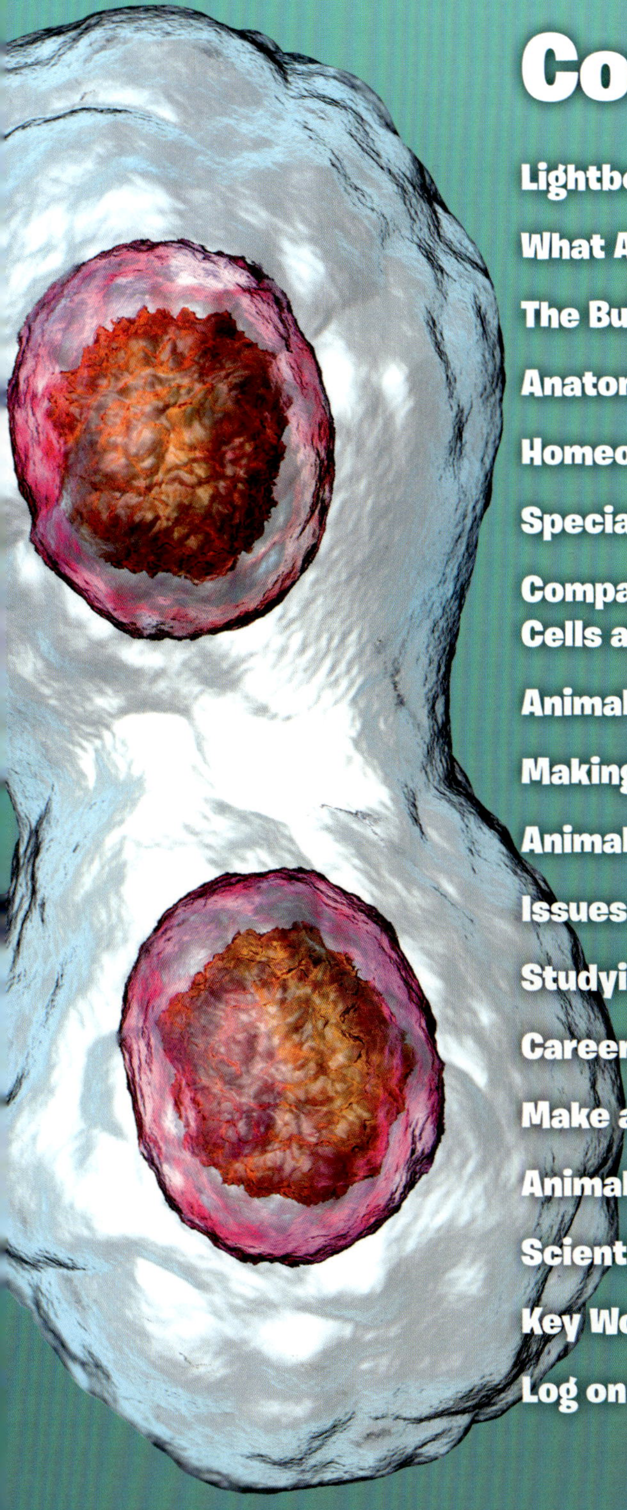

Contents

What Are Animals?

Animals make up at least 75 percent of all species on Earth. They range from enormous whales and elephants to tiny shrews and bats. Animals are multicellular organisms that must eat other organisms to survive, and that can move independently at some point in their life cycle.

The kingdom Animalia is divided into two categories called vertebrates and invertebrates. Vertebrates include all animals with a spinal column, or backbone. The five major classes of vertebrates are amphibians, birds, bony fish, mammals, and reptiles. Invertebrates, or animals without a backbone, include **arthropods**, **echinoderms**, jellyfish, **mollusks**, sponges, and worms.

Animals can also be classified according to their shape. They have bilateral **symmetry** when their left and right sides are symmetrical. All vertebrates are bilaterally symmetrical at some point in their life cycle. Jellyfish, seastars, and **hydra** are all animals with radial symmetry. This means that their body parts are arranged symmetrically around a central axis.

97% of all animal species are **invertebrates**.

Sponges are the only animals **without** a symmetrical body plan.

Of the more than **150 million** living things on Earth, about **75%** are **insects**.

There are approximately 5,000 different species of mammals on Earth.

The Building Blocks of Life

Like all living things, animals are made of cells. A cell is the smallest living unit of any life form. Each cell performs a specific function and has a finite, or limited, life cycle.

Some organisms, such as bacteria, are unicellular, meaning they consist of only one cell. Multicellular organisms such as animals are made of millions, or even trillions, of microscopic cells. Different kinds of animal cells include muscle cells that contract to produce movement, neurons that carry information, and fat cells that store energy.

Animal cells have their own life cycles. They pass on their genetic material when they reproduce. They consume energy and excrete waste. Eventually, they die. Just as multicellular organisms have organs that carry out functions, cells have organelles. These are structures within cells that carry out functions to keep the cell alive.

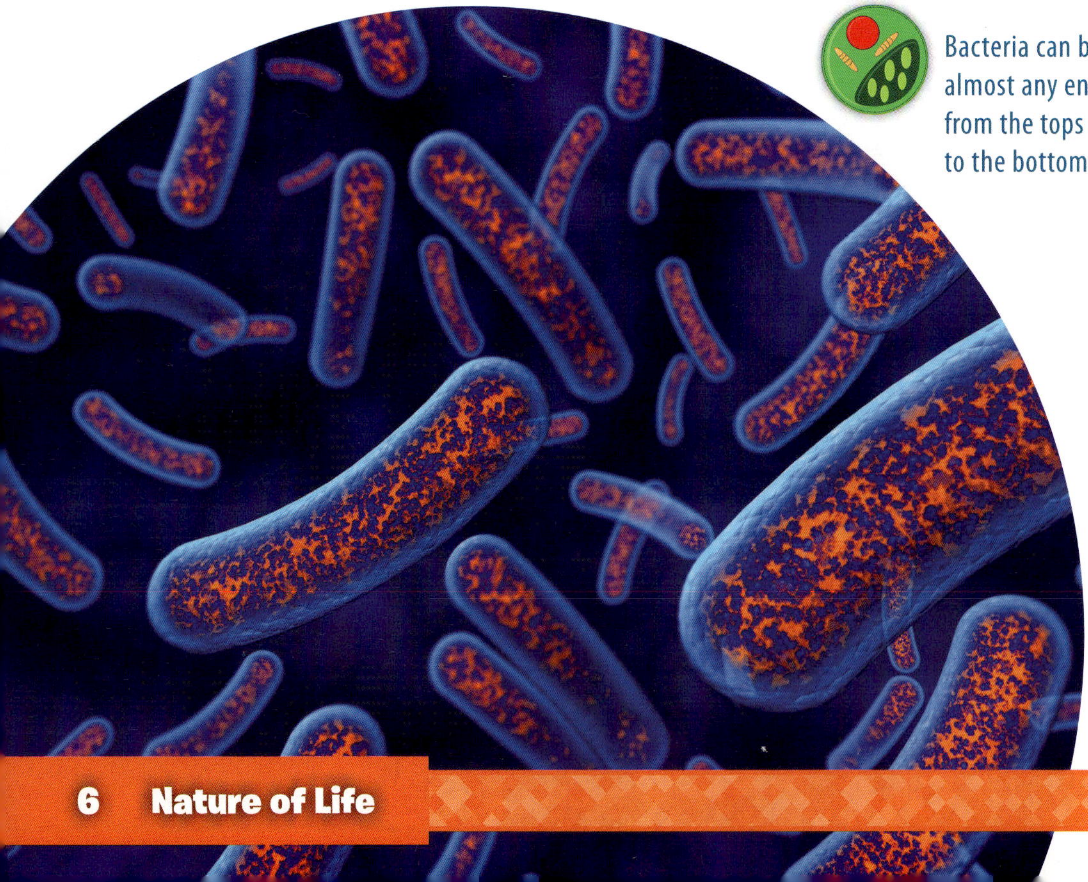

Bacteria can be found in almost any environment, from the tops of mountains to the bottom of the ocean.

Tissues, Organs, and Systems

Cells that perform similar functions organize to form tissue. There are four types of animal tissue. They are epithelial, connective, muscle, and nervous.

A collection of tissues that performs a function is called an organ. Skin, kidneys, stomachs, and brains are examples of animal organs. They work together in organ systems in **higher animals**.

There are many different organ systems. Humans have 10 organ systems. They are the circulatory, digestive, endocrine, urinary/excretory, integumentary, muscular, nervous, reproductive, respiratory, and skeletal systems.

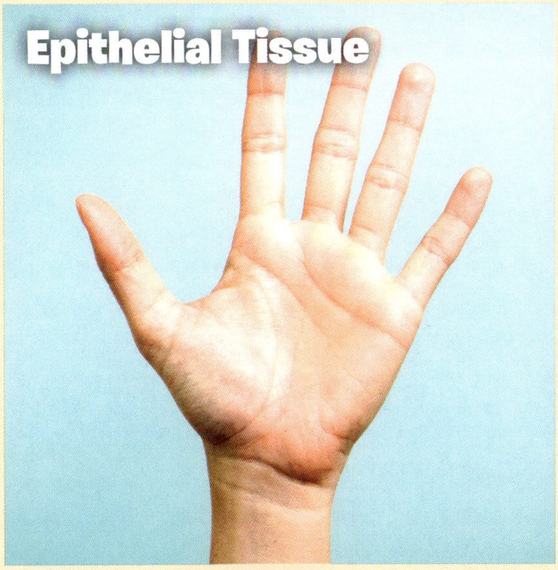

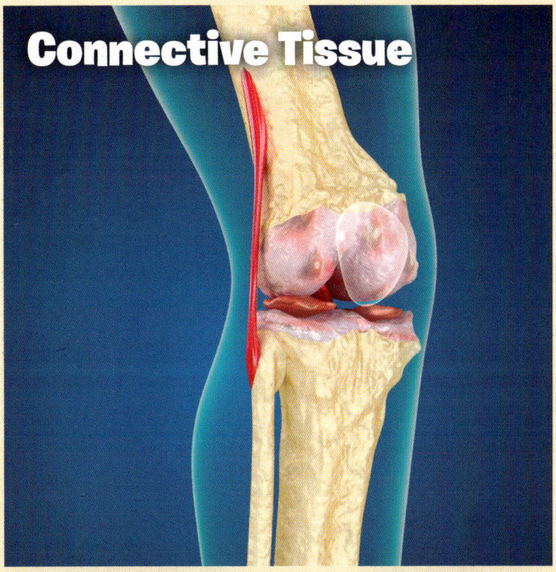

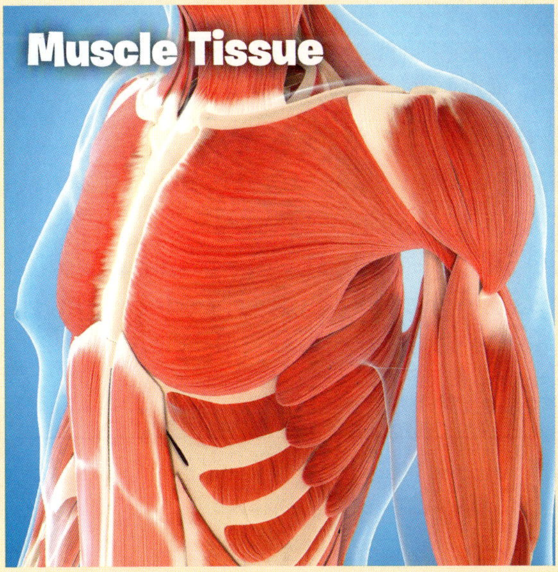

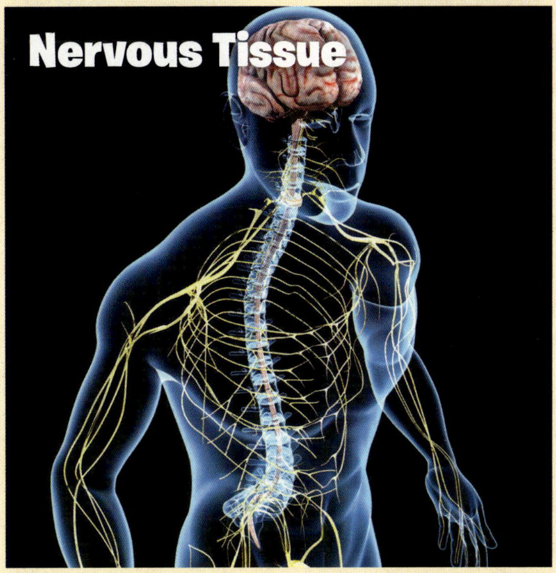

Anatomy of an Animal Cell

Every cell's job is to produce protein. Cells are like protein factories. Each organelle carries out a task related to protein synthesis.

1. **Cell membrane** The cell membrane is a thin layer of protein that surrounds the cell. Also called a plasma membrane, the cell membrane is semi-**permeable**, allowing substances such as glucose to move in and out of the cell. It represents the wall of the factory, with doors and loading docks to allow materials in and out.
2. **Cytoplasm** Cytoplasm is the thick fluid inside the cell membrane that contains the organelles. It is like the factory floor, where everything takes place.
3. **Smooth endoplasmic reticulum** The smooth endoplasmic reticulum is a site of lipid production.
4. **Rough endoplasmic reticulum** The rough endoplasmic reticulum is where the cell makes protein. Both the smooth and rough endoplasmic reticulum act as assembly lines in the factory.
5. **Golgi body** Golgi bodies are folded stacks of membranes. Animal cells have up to 20 Golgi bodies. The Golgi body gathers proteins to transport out of the cell. It is the factory's packing area, where goods are prepared for shipping.
6. **Lysosome** Lysosomes are sacs of digestive enzymes. They break down old organelles. They are the factory's housekeeping crew.
7. **Nucleus** The nucleus is the cell's largest organelle. It contains the chromosomes, nucleolus, nucleoplasm, and deoxyribonucleic acid (DNA). The nucleus controls many of the cell's functions and acts as the CEO's office.
8. **Nucleolus** The nucleolus is located inside the nucleus. It assembles ribosomes, which the cell uses to build proteins. It is the cell's human resources department that prepares workers for their jobs on the assembly line.
9. **Mitochondrion** Like nuclei, mitochondria are surrounded by two membranes. The inner layer is corrugated and the outer layer is smooth. The mitochondrion converts stored sugar into energy in the form of adenosine triphosphate (ATP). It is the generator that converts fuel into usable energy.
10. **Ribosome** Ribosomes assemble proteins in the cell. They are the factory's assembly line workers.
11. **Vesicle** Vesicles are liquid-filled membrane sacs that carry substances in and out of the cell. They are the factory's forklifts.

Eukaryotic Cell

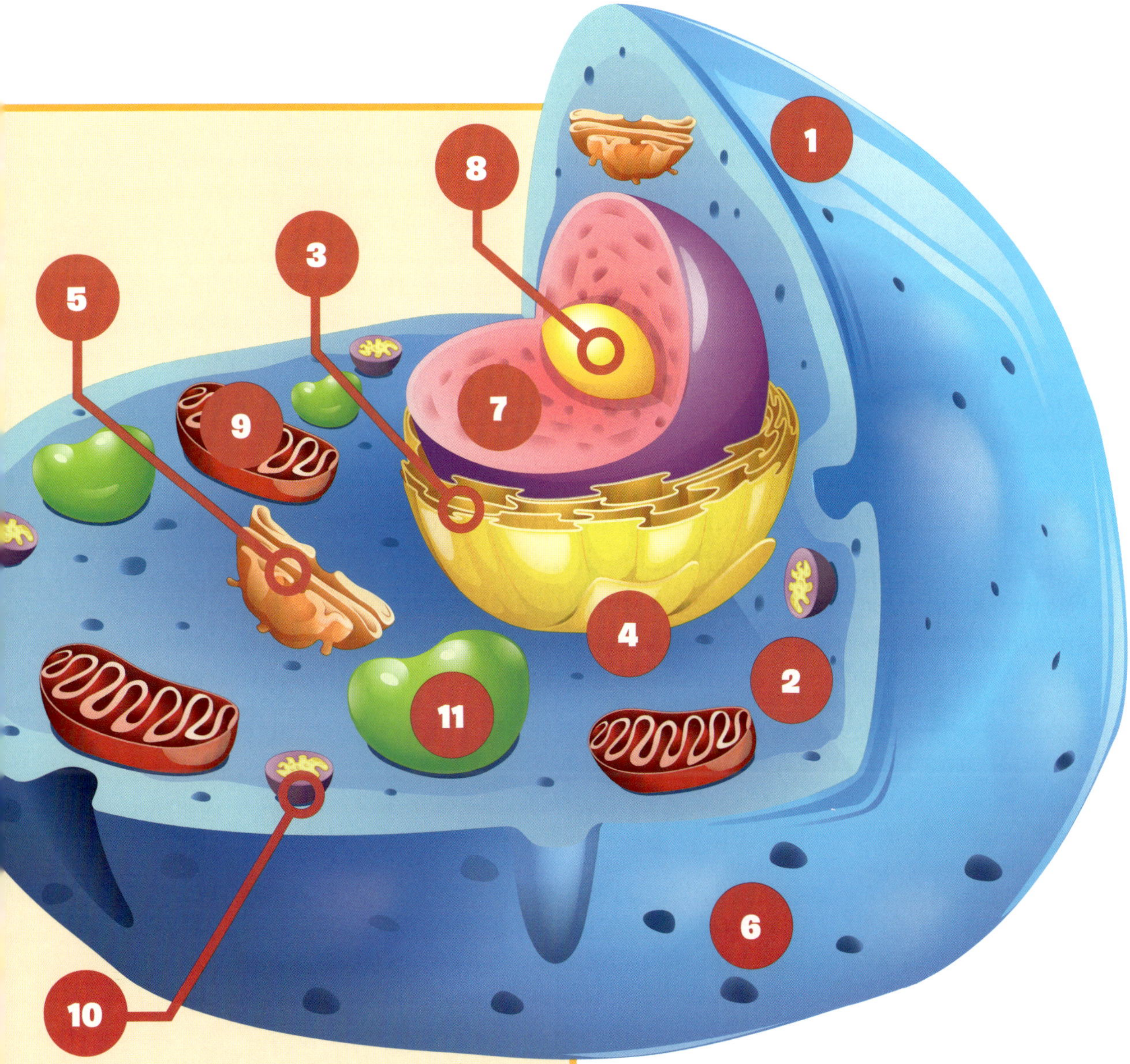

Homeostasis

Animals and animal cells need a constant internal environment in order to survive. They must maintain steady levels of water, salt, energy, heat, and waste. This steady state is called homeostasis.

One example of homeostasis is osmoregulation, the maintenance of steady water levels in animals and animal cells. At the cellular level, osmoregulation keeps cells in a balanced state, with neither too much nor too little water inside the cell membrane. When the concentration of salt and other electrolytes is higher inside the cell than in the surrounding fluid, water will pass from the surrounding fluid through the cell membrane and into the cell. Conversely, if the concentration of fluid inside the cell is lower than that of the fluid surrounding it, water will leave the cell. This movement of water across a membrane from a solution of low concentration to an area of high concentration is called osmosis.

In vertebrates, osmoregulation is monitored by a part of the brain called the hypothalamus. When an animal is dehydrated, this small structure signals the pituitary gland to release **antidiuretic** hormone (ADH). This hormone tells the kidneys to produce small amounts of urine with high salt concentrations. In this way, salt is removed from the bloodstream, while water loss is limited. At the same time, the hypothalamus stimulates feelings of thirst, prompting the animal to drink.

When there is too much water in the bloodstream, the hypothalamus detects the dilute solution. It signals the pituitary gland to produce less ADH. As a result, the kidneys produce a greater quantity of dilute urine. The body maintains salt and other electrolytes, and loses excess water.

Invertebrates such as insects have a waxy exoskeleton covering their bodies. This exoskeleton helps to prevent water loss.

Negative Feedback Loop

In vertebrates, the hypothalamus and kidneys work in a negative feedback loop to maintain osmoregulation.

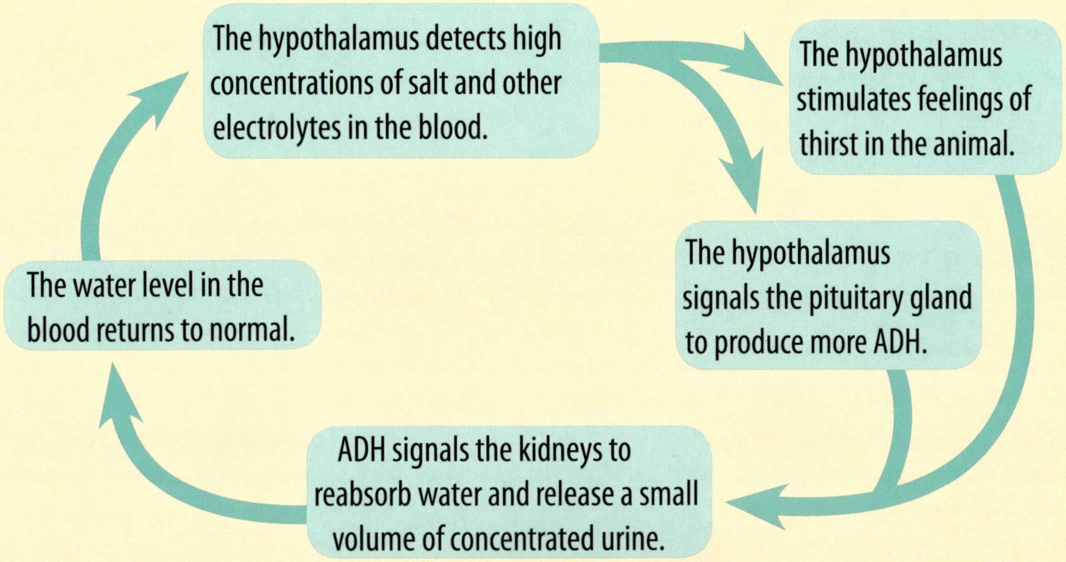

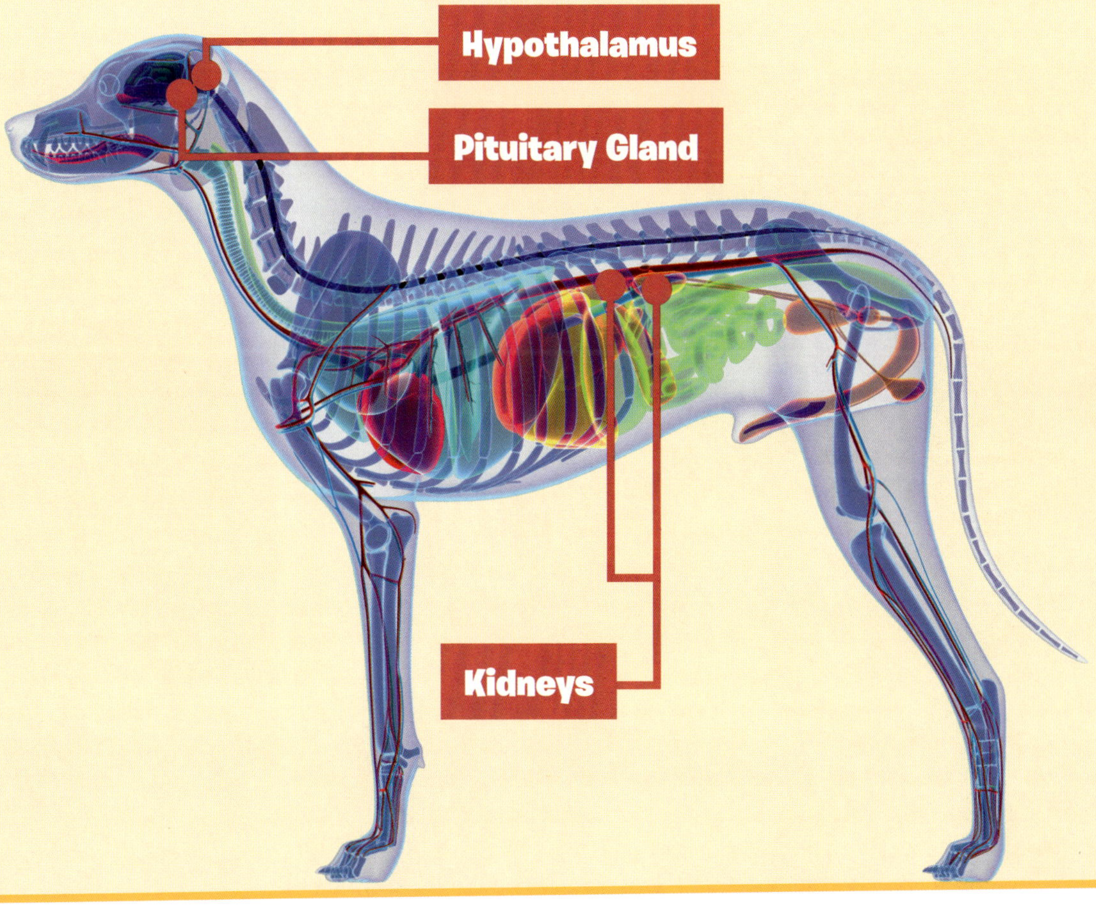

Specialized Cells

Cells develop different structures depending on their functions. Animal cells are specialized in many ways. They can perform a wide variety of functions.

Choanocytes are cells that line the inner bodies of sponges. Each cell has a structure called a **flagellum** to direct water into it, so that it can filter **plankton** and organic matter for the sponge to feed on. Choanocytes also play a part in sponge reproduction, catching sperm cells released by other sponges.

Red blood cells are specialized cells in vertebrates and some invertebrates. Hemoglobin, a protein that contains iron and binds to oxygen, gives red blood cells their color. Red blood cells carry oxygen from an animal's gills or lungs to other tissues and carry away carbon dioxide.

Even the yolks of North African ostrich eggs, the largest eggs of any animal, are single cells.

Neurons are the specialized cells of the nervous system. Each neuron consists of a cell body with long filaments, called axons and dendrites, extending from it. Axons are long, thin tubes that carry nerve impulses away from the cell. Dendrites are shorter structures that receive signals from other cells.

Egg cells are reproductive cells that develop into new organisms. Female eggs can be fertilized by male sperm. In organisms capable of parthenogenesis, a form of asexual reproduction, fertilization is activated without sperm.

Receptor cells are sensitive to stimuli. Animals can have receptors for pain, temperature, pressure, taste and smell, light, and other forms of energy. These receptor cells transmit information to an animal's nervous system.

Stem cells are undifferentiated cells, meaning they have no specialized function. They can, however, develop into specialized cells. In humans, stem cells are found in tissues such as bone marrow.

Fish gills contain filaments that expose red blood cells to the oxygen in water.

Signals travel through the human nervous system at speeds ranging from 1 to 268 miles (1.6 to 431 kilometers) per hour.

Comparing Animal Cells and Plant Cells

Cells make up all living things. Not only are they a part of animals, cells also make up plants, fungi, **protozoa**, mold, algae, and bacteria. Animals and plants are made of many specialized cells, tissues, and organs that work together to sustain life. Bacteria and protozoa are unicellular. Unicellular organisms may live alone or in colonies. Some species of fungi and algae are multicellular, while others are unicellular.

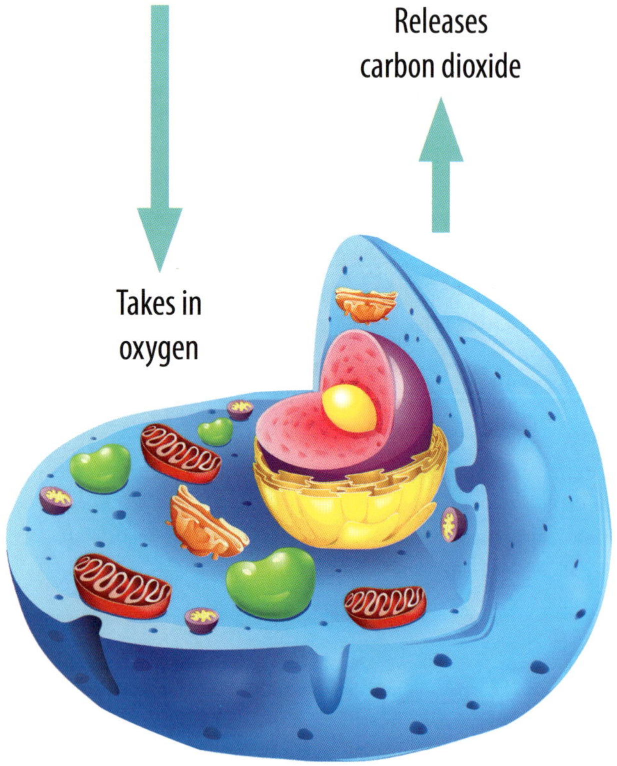

Animal Cell

Animal Cell
Animal cells consume oxygen and release carbon dioxide.
Animal cells have digestive organelles called lysosomes.
Animal cells have centrioles, which are protein structures that play a part in cell division.
Animal cells are surrounded by only a cell membrane.
Animal cells do not have chloroplasts.
Animal cells may or may not have small sacs called vacuoles.

Animals depend on plants for energy. Animals either ingest plant matter directly, or they eat other organisms that ingest plants. In contrast, plant cells make their own energy through a process called photosynthesis. Not surprisingly, animal and plant cells have different structures and functions.

As little as 10 percent of the energy plants get from the Sun is transferred to the animals that eat them.

Plant Cell
Plant cells create energy from carbon dioxide and sunlight. They release oxygen as a waste product.
Plant cells do not usually have lysosomes.
Plant cells do not have centrioles.
Plant cells are bound by a rigid cell wall that surrounds the cell membrane.
Some plant cells contain chloroplasts, which enable them to perform photosynthesis.
All plant cells have a large central vacuole.

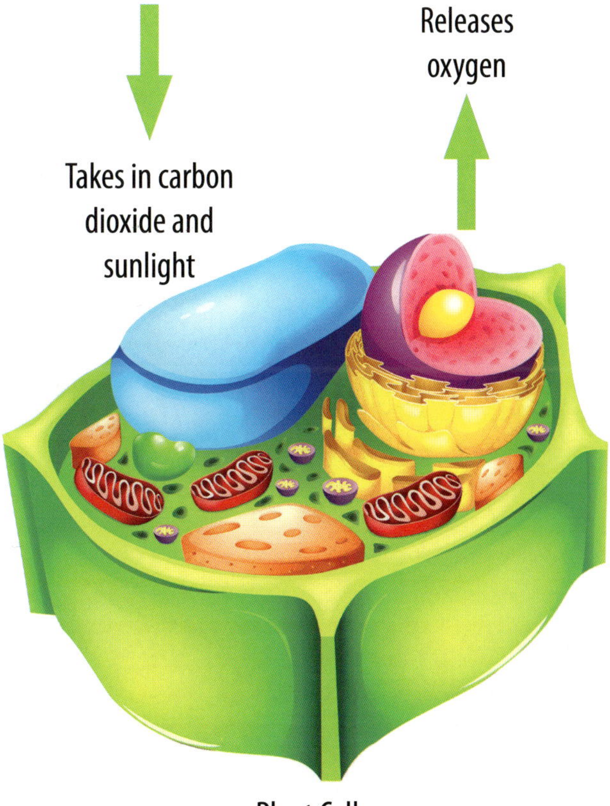

Plant Cell

Animal Adaptations

Animals live on every continent on Earth. Some live on land and others live in water. Animals have special features called adaptations that allow them to survive in their environments. Adaptations are often structural, but they exist at the cellular level, too.

Some insects that live in cold climates make their own **antifreeze**. The Upis beetle of Alaska can survive temperatures of –100° Fahrenheit (–73° Celsius), due to a sugar it produces. Attached to the outside of the beetle's cell membranes, the sugar molecules prevent ice crystals from entering the cells, keeping the contents of the cells from freezing.

Adaptations around the World

Alaska
Warm-blooded animals in cold Arctic climates can convert food energy to heat. Some species of fish and insects use chemical processes to keep their cells from freezing.

North America
The North American continent offers a variety of biomes and climates. This diversity is necessary to the life cycles of migratory insects, birds, and other animals.

Cephalopods such as octopuses, squids, and cuttlefish have special cells that help them hide from predators and attack prey. These animals can change the color of their skin with cells called chromatophores. Chromatophores lie under the surface of the skin. Each chromatophore contains a sac of pigment. When the cell is pulled by the network of muscles around it, the sac expands and the color appears brighter. When the sac contracts, the color fades.

Monarch butterflies migrate more than 2,500 miles (4,023 km) each year between Canada and Mexico. Monarch butterflies have clock cells in their antennae. These cells are photoreceptors that convey information about light conditions. The insects also use their eyes to judge the position of the Sun on the horizon. By coordinating information from their eyes and their antennae, the butterflies can use the Sun as a compass.

Australia
Camouflage is a common strategy for animals that live on the ocean floor. Some animals in this environment can change their color and texture to blend into their surroundings.

Making New Cells

Animal cells reproduce by dividing. Somatic cells divide by a process called mitosis. Reproductive cells divide by a process called meiosis.

Mitosis is the process by which a cell produces an exact copy of itself. Before the cell begins to divide, its chromosomes are scattered inside the nucleus. Chromosomes are X-shaped filaments of DNA that bear the cell's genetic information. As the cell prepares to divide, the chromosomes thicken. The nuclear membrane deteriorates, releasing the chromosomes. During this initial phase, the cell also makes a copy of an organelle called a centrosome. The two centrosomes move to opposite ends of the cell.

Next, a structure called a spindle grows between the two centrosomes. The spindle pulls the chromosomes into a row in the middle of the cell. The protein that binds the chromosome threads deteriorates and the paired threads divide into separate filaments.

The spindle structure pulls the newly divided chromosomes to opposite ends of the cell. New nuclei form around the newly divided chromosomes. Finally, the cytoplasm divides between the two new cells. The cell membrane pinches together between the new cells and separates them.

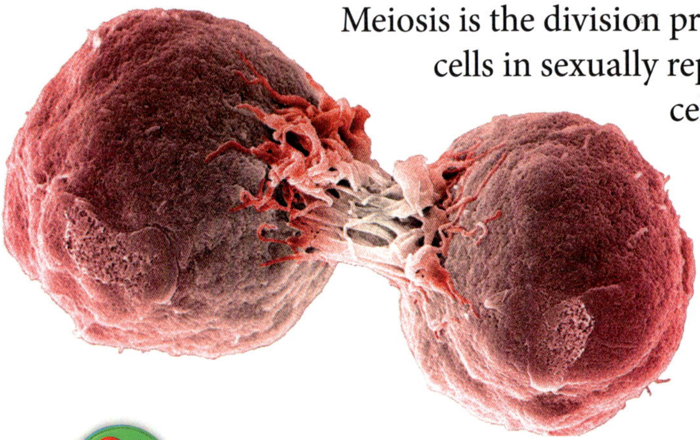

Meiosis is the division process of egg and sperm cells in sexually reproducing animals. These cells divide twice. In the first division, the two new cells have the same number of chromosomes as the original cell. The second phase of division results in four cells, each with half the number of chromosomes as the original cell.

Humans lose more than 30,000 skin cells each minute. The cells replace themselves by dividing constantly.

The red blood cells of mature mammals have no nuclei, so they cannot reproduce by division. These cells are produced in the bone marrow. Without a nucleus, red blood cells need less oxygen for their own metabolism, which enables them to carry more oxygen to other cells.

Animal Genetics

Most animals reproduce sexually. Sexual reproduction involves the union of an egg cell and a sperm cell. Each contains half of the animal's genetic information. In some species, male animals produce sperm cells and female animals produce egg cells.

Sexual reproduction can be internal or external. Internal fertilization is typical of animals that live on land. In this form of sexual reproduction, the sperm must be released in or near the reproductive tract of the female. External fertilization usually takes place among animals that live in water. Female animals lay their eggs, and male animals deposit sperm on or near the eggs.

Animals such as tapeworms are hermaphroditic, meaning they are able to produce both egg and sperm cells. They can fertilize their own eggs or those of other organisms. Some fish, lizards, birds, and invertebrates can produce eggs that develop into embryos without being fertilized.

Fertilization

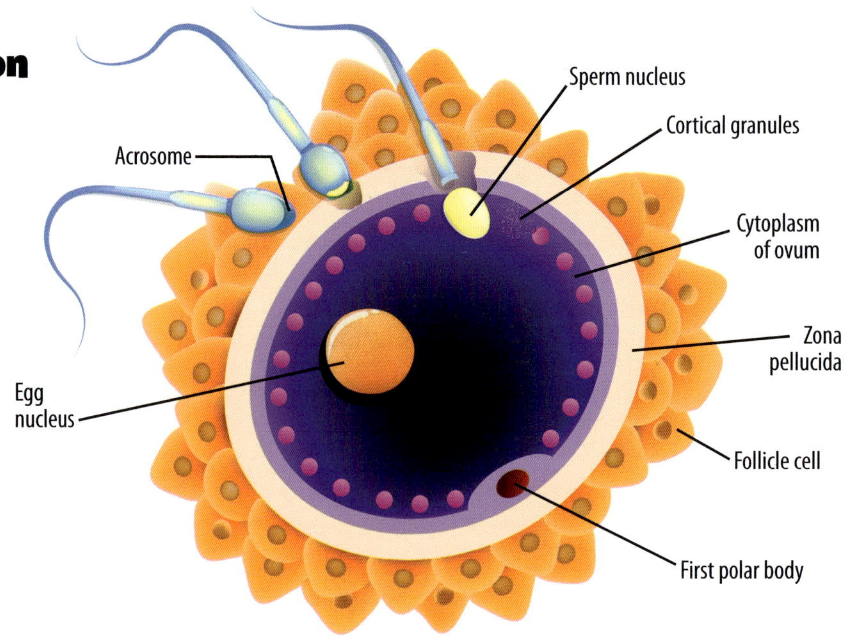

The nucleus from the egg combines with the nucleus from the sperm. This creates the first nucleus for a new organism.

Animals inherit traits from their parents. These traits vary among individuals and are encoded in the DNA that the animal's parents pass down. Natural selection is the process by which individuals that are better suited to their environment are more likely to survive and reproduce. Sometimes, a population of animals develops similar traits that enable them to thrive in their environment. Over time, animals without the new traits do not survive to pass on their genes.

Predicting Coat Color

A Punnett square shows how traits will be distributed when animals are bred.

In the 1850s, Gregor Mendel, an Austrian monk, experimented with pea plants. He wondered how plants that bore green peas could result from parent plants that bore yellow peas.

The Punnett square diagram, developed by Reginald Punnett in 1905, solves Mendel's query by showing how dominant and recessive traits work.

In this diagram, a Punnett square is used to predict the coat colors of cattle. Both parents have black coats, but carry the recessive gene for red coloring. E represents the dominant trait of black coat coloration, and e represents the recessive trait of red coat coloration.

The diagram shows that 75 percent of the offspring will have black coats, and 25 percent will have red coats.

	E	**e**
E	EE	Ee
e	Ee	ee

Issues and Oddities

Sometimes, cells reproduce at an excessive rate. This is called cancer. Over time, these rapidly produced cells may form a mass, or tumor, in the tissue. Cancer cells can spread via an animal's circulatory system, invading tissues far from where the cancer started. When cells divide, sections of DNA called genes may be copied incorrectly. These changes are called mutations. Cancer is one type of mutation.

Mutations in egg and sperm cells are heritable, meaning they can be passed on to an animal's offspring for generations. Somatic mutations occur in body cells. They affect only the individual organism and are not inherited by its offspring.

Hereditary diseases are the result of genetic mutations. Cystic fibrosis and Huntington disease are examples of hereditary diseases in humans. In dogs, hip dysplasia, a condition in which the femur fits too loosely in the hip socket, is a hereditary disorder.

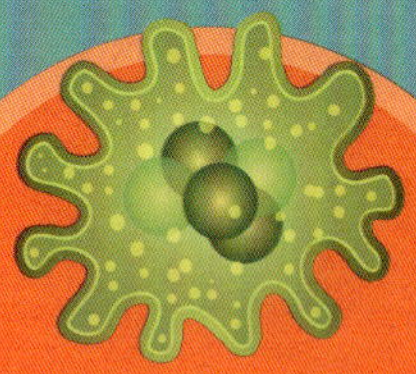

Mutations can be **positive or negative.**

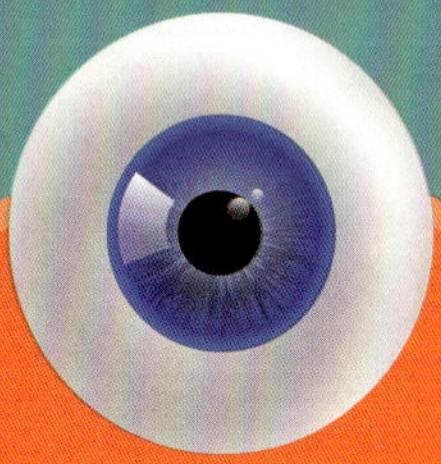

Blue eyes in humans result from a genetic mutation that occurred **6,000 to 10,000** years ago. All blue-eyed people share a **single common ancestor**.

A domestic goldfish's **brightly colored scales** are the result of a genetic mutation. In nature, these fish are **gray** so they are not seen by predators.

Adaptations are mutations that allow populations of animals to thrive in adverse conditions. Some species of kingsnake, for example, have a pattern of stripes on their skin that resembles that of a coral snake. This adaptation protects the venomless kingsnake. Since predators mistake it for its poisonous lookalike, they do not attack it.

Mutations can occur spontaneously. They can also result from an organism's exposure to **mutagens**. Radiation and pollutants, such as pesticides, are common mutagens.

Mutations in pets and livestock can result in traits that are valued by humans. Artificial selection, or selective breeding, is the practice of breeding animals for such desired traits. Animals that possess the desired traits are bred with each other, while animals without these traits are not allowed to reproduce. Hypoallergenic cats, flat-faced bulldogs, and high-yielding dairy cattle are all the results of selective breeding.

The Holstein dairy cattle breed is approximately 2,000 years old. One Holstein can produce more than 72,000 pounds (32,659 kilograms) of milk each year.

Studying Animal Cells

How do cells communicate? How do cancer cells respond to different treatments? How did life begin? Scientists who study cells ask big questions. They have made important discoveries in medicine and genetics.

1665 Robert Hooke is the first person to observe plant cells through a microscope. His scientific findings are published in a book called *Micrographia.*

1855 Rudolf Virchow, a German doctor, states that all cells come from existing cells. This statement completes the cell theory first proposed by Schwann and Schleiden.

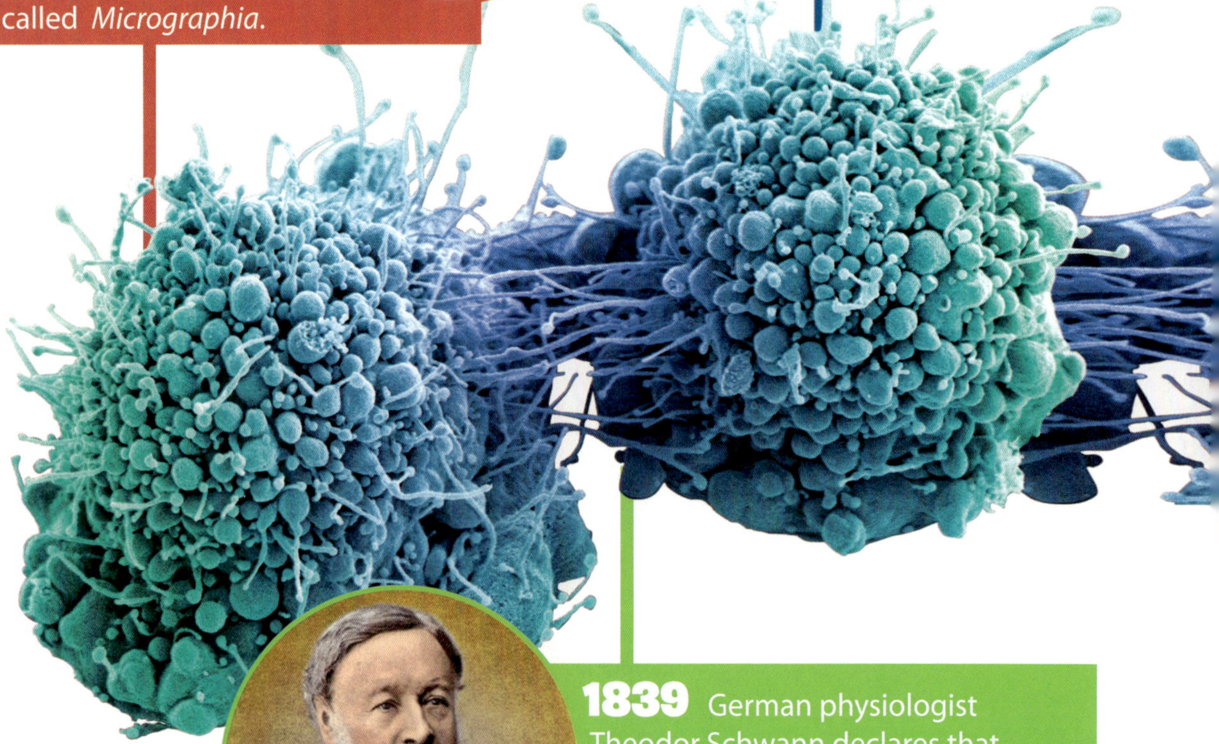

1839 German physiologist Theodor Schwann declares that cells are the building blocks of all animal life. In the same year, German botanist Matthias Schleiden makes the same observation about plant cells.

1882 German biologist Walther Flemming explains in detail the process of cell division. He is the founder of cytogenetics, the branch of genetics that focuses on the structures and functions of chromosomes.

Today

John Tyler Bonner, biologist

John Tyler Bonner began his studies at Harvard University in 1937. After serving in the Air Corps from 1942 to 1946, he returned to Harvard to complete his PhD in 1947. Soon after, Bonner joined the faculty of Princeton University, where he currently holds the title of George M. Moffett Professor of Biology, Emeritus in the Department of Ecology and Evolutionary Biology. Recognized as one of the world's most important biologists, Bonner has used simple organisms called slime molds to better understand complex multicellular animals.

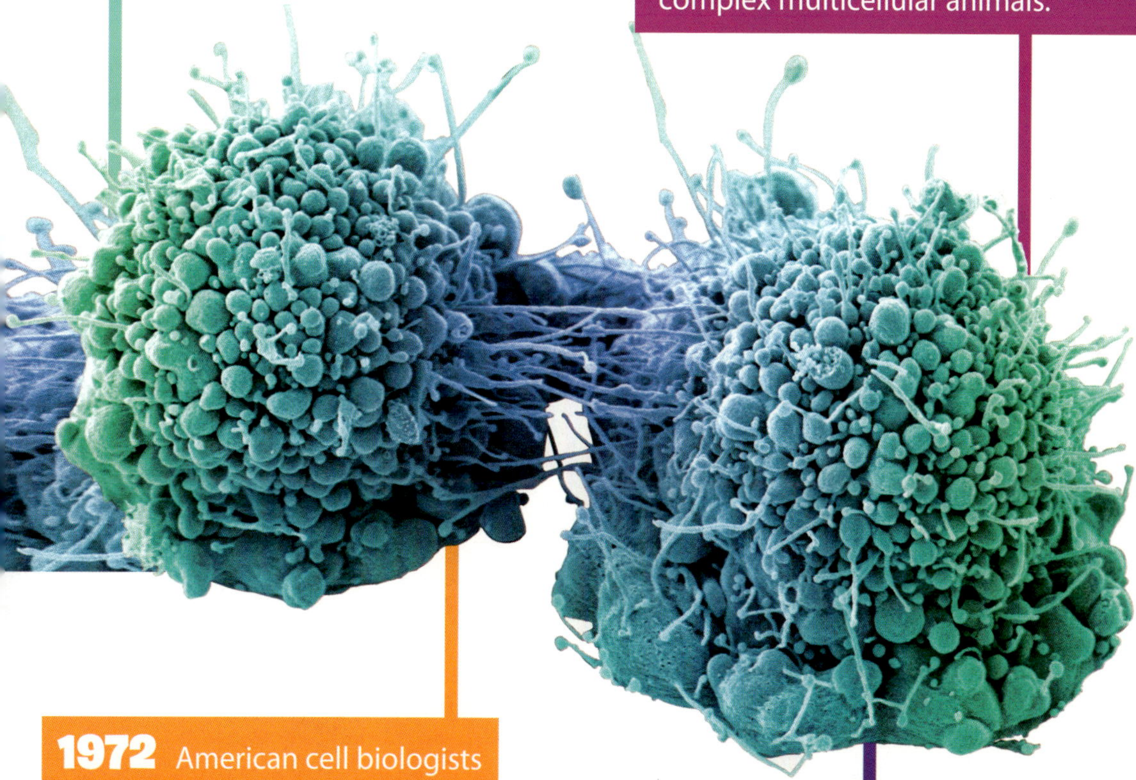

1972 American cell biologists Seymour J. Singer and Garth L. Nicolson propose the current model of the cell membrane. They describe it as a double layer of lipids with proteins attached to and embedded in the layers.

2003 The Human Genome Project is completed. Started in 1990, the project set out to sequence all 20,000 to 25,000 human genes and make the sequences publicly available for research applications.

Careers

The study of animal cells can lead to various careers in medicine or biology. People who study animal cells might work in a laboratory, classroom, or office. Their research can contribute to fighting diseases or identifying new animal species. Knowledge of genetics can help people plan their families or cope with hereditary diseases.

Molecular Biologist

Molecular biology is the study of the chemical processes that take place within cells. Molecular biologists are knowledgeable in cell physiology, genetics, physics, and chemistry. Through their work, they may develop new ways to fight and prevent diseases, or make important discoveries about genes and heredity. Using sophisticated lab equipment, these scientists may work on projects that involve cloning or drug testing. Some teach post-secondary students. Molecular biologists can work in universities as teachers and researchers, as forensic scientists in the public service, or as product developers in private biotechnology firms. A career as a molecular biologist requires a bachelor's, master's, or doctoral degree.

Cell Culture Technician

Cell culture technicians grow cell cultures from tissue samples or commercial sources. They perform tasks such as preparing and maintaining cell growing mediums. Cell culture technicians are often experts in microscopy. Their work may be part of medical, pharmaceutical, or

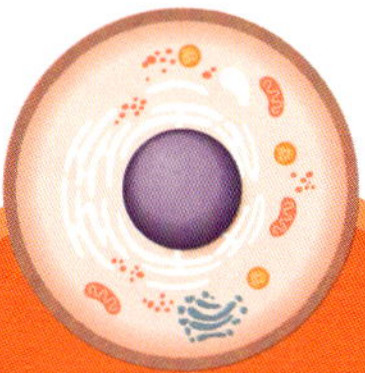

Cytology, the study of cells, became a distinct field of biology in the **late 19th century**.

Forensic scientists analyze tissue samples to find DNA evidence and help **solve crimes**.

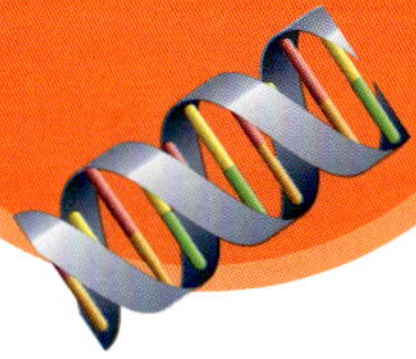

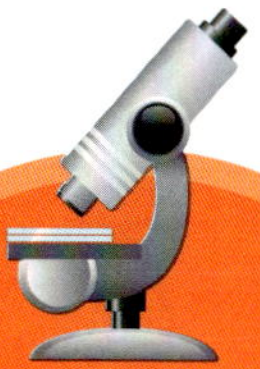

The term **molecular biology** was coined in the 1930s by a mathematician named Warren Weaver.

biotechnical research. Cell culture technicians work in laboratories with other scientists. They conduct experiments, write reports, and may also teach university students. Most cell culture technicians work for universities or private firms. This career requires a bachelor's or master's degree in biology.

Genetic Counselor

Genetic counselors advise people about hereditary illnesses. They have expert knowledge of medical genetics and statistical analysis, as well as high emotional intelligence and strong communication skills. Genetic counselors work closely with patients and families to help them assess the risk of passing on hereditary diseases. They also diagnose hereditary conditions and may work with **assisted reproductive technologies**. Genetic counselors work in teams with other health care professionals. Many of these professionals begin their careers in healthcare or research. They work in hospitals, private clinics, and universities. Genetic counselors require a master's degree in the field.

Make a 3D Model of an Animal Cell

Materials

Modeling clay in different colors

White glue

Pencil

Strips of paper

Toothpicks

Instructions

1. Mold a large piece of clay into a shallow dish shape. It should be about 6 inches (15 centimeters) in diameter.
2. Mold different colors of clay into organelles. Be sure to include the nucleus, Golgi body, lysosomes, mitochondrion, smooth endoplasmic reticulum, and rough endoplasmic reticulum.
3. Write the names of the organelles on the strips of paper. Glue each label to a toothpick and stick it into the corresponding organelle.
4. Pour a shallow layer of glue into your model, surrounding all of the organelles. This represents the cytoplasm.
5. Let your model dry overnight. You have now created your own 3D model of an animal cell.

Animal Cells Quiz

1. What is bilateral symmetry?

2. What is a group of cells that perform similar functions called?

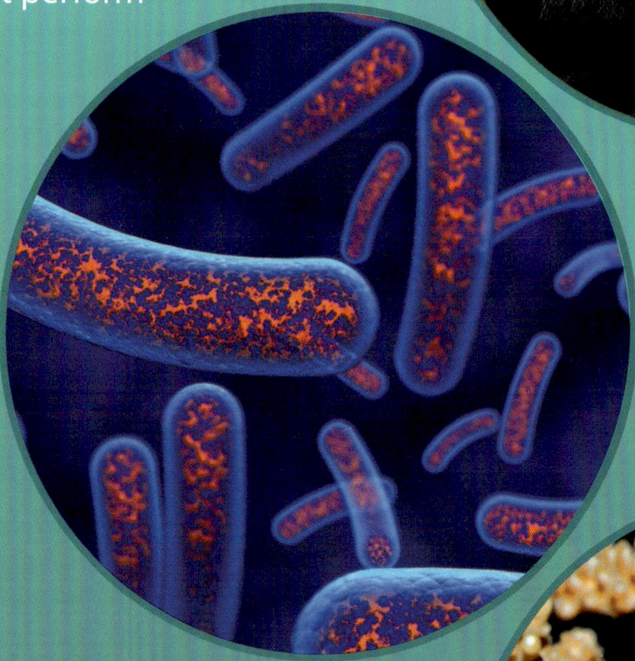

3. Which organelle is the site of lipid production?

4. What is cytoplasm?

5. What organelles do animal cells have that plant cells usually lack?

6. What are the female and male reproductive cells in some species?

7. What is osmoregulation?

8. What is a mutation?

9. What is an adaptation?

10. What is the difference between mitosis and meiosis?

Answers: 1. A body shape in which the left and right sides are symmetrical 2. Tissue 3. The smooth endoplasmic reticulum 4. The thick fluid inside the cell membrane 5. Lysosome, centriole 6. Egg cells and sperm cells 7. The maintenance of a steady level of water in a cell or organism 8. A change caused by an incorrect copy of a gene 9. A mutation that helps a population of organisms to survive in different environments 10. Mitosis is the division of somatic cells and meiosis is the division of reproductive cells

Scientific Terms

The study and practice of science comes with its own language. This list will introduce you to some common scientific terms. Understanding the meaning of these terms will allow you to discuss scientific concepts more effectively with others.

adaptations: inherited changes that allow organisms to survive in adverse conditions

carbon dioxide: a gaseous molecule consisting of one carbon atom and two oxygen atoms

cell: the smallest structure in the body of a living thing

chromosomes: threads of DNA found in the nucleus of cells

DNA: deoxyribonucleic acid, the component in every cell in which genetic information is encoded

dominant: a gene that, when paired with a recessive copy of the same gene, will be expressed in the offspring

gene: a segment of DNA that codes for different traits in an individual

heredity: the genetic transmission of characteristics from parent to offspring

microscopic: something that is too small to see with the naked eye and can only be seen with the aid of a microscope

migrate: to move seasonally from one habitat to another

multicellular: made of many cells working together in tissues and organs

nucleus: the part of a cell that controls its actions and contains genetic information

organism: a living thing

osmosis: the movement of water across a barrier from an area of high concentration to an area of lower concentration

oxygen: a molecule that is essential to human, animal, and plant life

predator: an organism that consumes animals for food

protein: complex molecules made from amino acids that carry out numerous functions in cells

recessive: a gene that, when paired with a dominant copy of the same gene, will not be expressed in the offspring

species: a group of organisms with similar traits that are able to produce young

traits: variations that can be passed down from one organism to the next generation

Key Words

antidiuretic: suppressing the production of urine

antifreeze: a substance that lowers the freezing point of a liquid

arthropods: a phylum of animals with segmented bodies and jointed legs

assisted reproductive technologies: medical interventions to bring about pregnancy

camouflage: coloring, and sometimes texture, that allows an animal to blend in with its surroundings

echinoderms: a phylum of marine invertebrates with radially symmetrical bodies

flagellum: a tail-like structure

higher animals: animals with advanced characteristics, such as mammals

hydra: an aquatic animal with a tube-like body

mollusks: a phylum of marine animals with soft, unsegmented bodies; many mollusks have hard shells

mutagens: things that cause cells to mutate

permeable: allowing some substances to pass through

plankton: small water-dwelling organisms that are a food source for marine and aquatic animals

protozoa: single-celled organisms other than bacteria

symmetry: the even arrangement of parts along a line or around a center

Index

LIGHTBOX

SUPPLEMENTARY RESOURCES

Click on the plus icon found in the bottom left corner of each spread to open additional teacher resources.

- Download and print the book's quizzes and activities
- Access curriculum correlations
- Explore additional web applications that enhance the Lightbox experience

LIGHTBOX DIGITAL TITLES
Packed full of integrated media

VIDEOS

INTERACTIVE MAPS

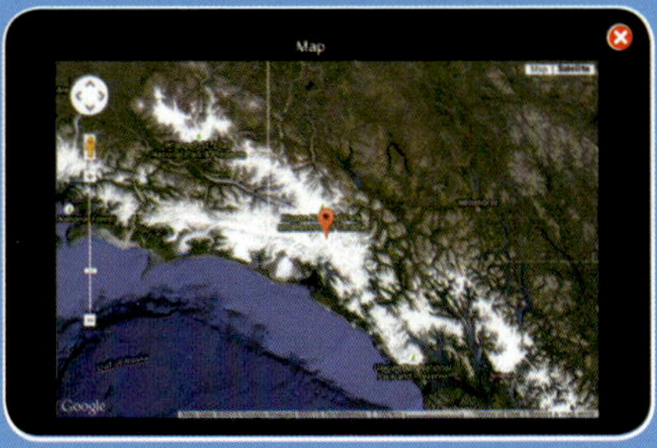

WEBLINKS

SLIDESHOWS

QUIZZES

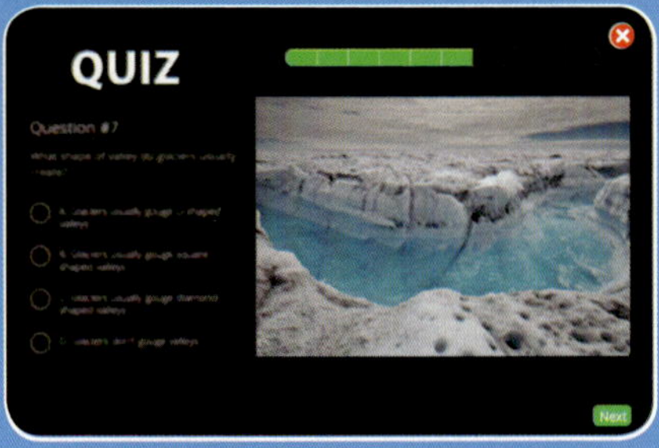

OPTIMIZED FOR

- ✓ TABLETS
- ✓ WHITEBOARDS
- ✓ COMPUTERS
- ✓ AND MUCH MORE!

Published by Smartbook Media Inc.
350 5th Avenue, 59th Floor
New York, NY 10118
Website: www.openlightbox.com

Library of Congress Cataloging-in-Publication Data

Names: Lomberg, Michelle.
Title: Animal cells / Michelle Lomberg.
Description: New York, NY : Smartbook Media, Inc., [2017] | Series: Nature of life | Includes index.
Identifiers: LCCN 2016020198 (print) | LCCN 2016021647 (ebook) | ISBN 9781510511842 (hard cover : alk. paper) | ISBN 9781510511859 (Multi-user ebk.)
Subjects: LCSH: Cells--Juvenile literature. | Life (Biology)--Juvenile literature.
Classification: LCC QH582.5 .L66 2017 (print) | LCC QH582.5 (ebook) | DDC 571.6--dc23
LC record available at https://lccn.loc.gov/2016020198

Printed in Brainerd, Minnesota, United States
1 2 3 4 5 6 7 8 9 0 20 19 18 17 16

072016
070816

Editor: Katie Gillespie
Designer: Mandy Christiansen

Photo Credits
Every reasonable effort has been made to trace ownership and to obtain permission to reprint copyright material. The publisher would be pleased to have any errors or omissions brought to its attention so that they may be corrected in subsequent printings. The publisher acknowledges Getty Images, iStock, Alamy, Minden, Shutterstock, and Dreamstime as its primary image suppliers for this title.